THE
MEDITERRANEAN
AND ITS PEOPLE

David Flint

Wayland

PEOPLE
· AND PLACES ·

The Alps and their People

The Amazon Rainforest and its People

The Arctic and its People

The Ganges Delta and its People

The Islands of the Pacific Rim and their People

The Mediterranean and its People

The Prairies and their People

The Sahara and its People

Book editor: Judy Martin
Series editors: Cally Chambers and Paul Mason
Designer: Mark Whitchurch

Consultant: Dr Tony Binns, lecturer in geography at the University of Sussex

First published in 1994 by Wayland (Publishers) Ltd
61 Western Road, Hove, East Sussex, BN3 1JD, England

© Copyright 1994 Wayland (Publishers) Ltd

British Library Cataloguing in Publication Data
Flint, David
   Mediterranean and Its People. – (People
   & Places Series)
   I. Title II. Series
   304.209182

ISBN 0-7502-0489-3

Typeset by Dorchester Typesetting Group Ltd
Printed and bound in Italy by G. Canale & C.S.p.A.

Acknowledgements
The publishers would like to thank the following for allowing their photographs to be reproduced in this book:
Camera Press 32 left and right, 33 (Alfa); J. Allan Cash 17, 27, 30; Cephas 6 top (Sand Hambrook), 22 bottom (Mick Rock); Chapel Studios 28 (Oliver Cockell), 42, 43 top and bottom (Zul Mukhida); C. M. Dixon 34; Ecoscene 10 (R. A. Beatty), 24 top (Wilkinson), 38 (Anthony King/Medimage), 41 (Groves); Eye Ubiquitous cover (Mike Southern), title page (Julia Waterlow), contents page (Hugh Rooney), 5 (Mike Southern), 6 right (Helene Rogers/Trip), 8 both 12 (Simon Arnold), 14 and 15 (Julia Waterlow), 18, 24, 26 (Ken Howard), 37, 45 (BB Pictures); G.S.F. Picture Library 19 (R. Gordon), 20 bottom; Denis Hughes-Gilbey 20 top; Panos Pictures 36 (Penny Tweedie); Tony Stone Worldwide 25 (Alan Smith), 31 (David Hanson); Topham Picture Library 21 (Judyth Platt), 23; Wayland Picture Library 16, 39 (Jimmy Holmes), 40; Worldsat Productions/NRSC/Science Photo Library 4 top.
Artwork by Peter Bull (4 bottom, 22 top) and Tony Smith (9, 11 and 27).

Cover: The picturesque fishing village of Vernazza in Cinqueterre, on the west coast of Italy. The unspoiled view evokes a flavour of the Mediterranean's history and traditional ways of life.

Title page: Workers pack the orange harvest into crates in the Nile delta in Egypt.

Contents page: This view down to Amalfi, in south-western Italy, shows how the steep hillsides are cut into narrow, stepped terraces of bare soil for cultivation of food crops.

C O N T E N T S

THE·NATURAL·ENVIRONMENT

*T*he Mediterranean Sea is the sixth largest area of water in the world. It covers 2,509,000 square kilometres, which is about ten times the size of the UK. This book looks at both the sea and the land around its shores.

Historically, this is one of the most important parts of the world. From the northern Mediterranean, great civilizations based on Greece and Rome grew and spread to conquer much of the then-known world. The sea has enabled the people of the Mediterranean to travel directly between the nations around its shores and for centuries has been a vital trade route connecting the Middle East with Europe.

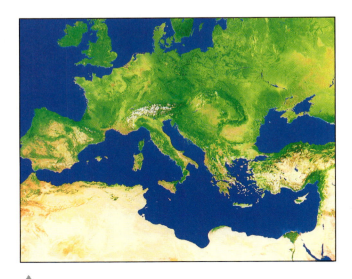

▲
A satellite view showing areas of dense (green) and sparse (brown) vegetation.

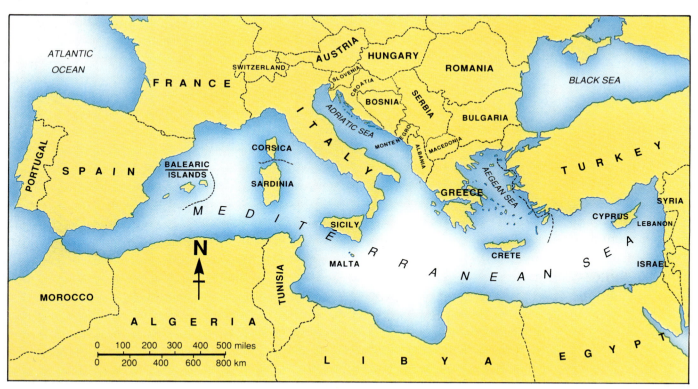

▲
The Mediterranean region. Bosnia, Croatia and Serbia are proposed divisions of former Yugoslavia.

A fisherman mending his nets on the Greek island of Samos. Many people make a modest living by fishing from small boats off the coasts and islands of the Mediterranean.

In addition, the Mediterranean has a distinctive climate. This generated a natural system of plants, birds and animals adapted to the region, which lived in harmony for hundreds of years. During the twentieth century, all this has been changing, because the balance between people, plants, animals, soil and sea has been gradually altered by human activity. Developments in farming, industry and tourism have produced economic benefits for the Mediterranean people, but often at the cost of damage to their environment. More recently, growing concern for the region's ecology has led to new development projects intended to preserve or restore the environmental balance.

A VERY SPECIAL SEA

The Mediterranean Sea is divided by Italy and Malta into a western and an eastern basin. Water enters the Mediterranean on the western side from the Atlantic Ocean via the Strait of Gibraltar. Here the water is only 350 m deep, so the flow of water from the Atlantic is limited. Inside the western basin, the sea floor drops steeply to over 2,000 m. Near the island of Sicily, the southernmost part of Italy, the water is again only 300 m deep, but in the eastern basin it plunges to a depth of over 3,000 m.

The water in the Mediterranean is a clear, deep blue, which shows it is low in plant and animal food. This is not the result of pollution, but is due to the salty nature of the deep water. Water evaporates rapidly from the surface of the sea, leaving saltier, heavier water that sinks to the lower levels.

Around the edge of the sea there is a narrow continental shelf of shallower water. Here plankton, those minute plants and animals of the sea, grow well, so here too are found the fish, squid and jellyfish that feed on the plankton. Because the continental shelf is narrow, the Mediterranean has relatively little plankton for its size and relatively few types of fish. The main varieties are tuna, swordfish and sardines, which are caught for food, together with some dolphins and porpoises. Fishing has always been an important feature of Mediterranean life, but because of the limited fish stocks, there is no large-scale fishing industry.

▲
The strange landscape of the Sahara desert, over which the sirocco wind blows northwards.

▶

Waves whipped up by the sirocco on the Mediterranean coast of Morocco.

WARM WINTERS, DRY SUMMERS

Around the Mediterranean coast, the weather is hot and dry in summer. In the north, winters are warm but wet, with short, heavy rainstorms. There is low rainfall throughout the year in the south-eastern area of the Mediterranean; parts of Egypt and Libya, for example, have true desert conditions.

The hot, dry summers have made the Mediterranean a paradise for holiday-makers from all over the world. For local people, the continuous fine weather has disadvantages. The summer sun ripens crops such as grapes, olives, oranges and lemons, but the heat dries up the grasslands. In most places there is not enough summer pasture for sheep and goats. Fires that break out in forest and scrubland areas spread easily through the dry vegetation.

Local winds also affect life in the Mediterranean. In spring, cold north winds blow down the Alps. In France, the mistral wind in the Rhône valley reaches speeds of up to

130 km/hr. Farmhouses in the valley are built with no doors or windows on the northern side and, to break the force of the cold mistral, rows of poplar or cypress trees are planted. A similar cold north wind called the bora affects the countries bordering the Adriatic region of the Mediterranean coast.

Winds from the south bring hot, dry conditions. The sirocco originates in north Africa and blows across the Mediterranean into southern Europe. Olive trees and vines can be damaged by the heat, and also by the red dust from the Sahara desert that is carried on the wind.

Typical north-eastern Mediterranean landscape, in southern Greece.

THE MEDITERRANEAN LANDSCAPE

The original vegetation of the lands around the northern Mediterranean was forest, which covered the region 8,000 years ago. Holm oak and cork oak were the main native trees in western areas. Further east, the Aleppo pine, cypress and cedar trees flourished. These trees are adapted to withstand long summer drought. They have extended roots that can draw water up from deep underground and thick bark that helps to cut down the loss of water by evaporation, They have either few leaves or narrow needle-shaped leaves, both of which reduce water loss.

Forest fires devastate the landscape and may threaten homes and tourist sites.

Forest vegetation is tall and gives dense cover in the tree canopies and at lower levels where shrubs and grasses grow

Maquis is lower-growing, up to 2 m in height, and consists of shrubs, flowering plants and grasses

Garrigue is sparse vegetation, with small plants that can withstand dry conditions growing on stony soil and among rocks

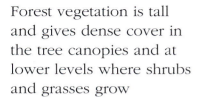

The different types of vegetation seen in the northern Mediterranean countries.

The forest floor was covered with grasses that would die back to their roots in the summer drought, then grow again when the winter rains returned. Between the grass and trees grew flowering plants and shrubs also adapted to survive the Mediterranean climate. Bushes like rosemary have tightly rolled leaves that cut down the amount of surface evaporation. The leaves of the hairy lupin are covered in felted hairs, which help to keep water in the plant. Bee and spider orchids also thrived among the tangle of grasses and bushes.

Very little of the forest vegetation remains. Over time, many trees have been felled for fuel and to provide timber for buildings and ships. Sheep and goats have eaten away the new shoots of saplings, and forest fires have destroyed many mature trees. In many places, the forest has given way to maquis, a rough mass of tall shrubs such as the mastic tree, myrtle and cistus. These grow up to 2 m high, much shorter than the forest trees, and in summer become a dry tangle of brittle stems and leaves. In spring, the maquis is made colourful by yellow flowering broom, and the white or pink blooms of tree heather and cistus.

Garrigue is a type of vegetation that replaces the original forest in drier areas. In the garrigue, bushes grow to only 0.5 m high, with much sandy soil and rock exposed between the plants. The kermes oak survives, with its dark green, leathery, holly-like leaves; a waxy covering on the leaves helps to keep water in the tree. There are over 200 types of scented plants in the garrigue areas of Greece, including herbs such as thyme, sage and rosemary.

▲
Grazing goats have contributed to the formation of low-level maquis vegetation.

MEDITERRANEAN WILDLIFE
The Mediterranean region is home to a wide range of insects, reptiles, animals and birds. These include lizards like the wall lizard and the chameleon, best known for its rapid colour changes. There are snakes, frogs and newts, scorpions and beetles, such as the stag beetle.

Insects live in and feed on the vegetation. Crickets and cicadas make a lot of noise, but mosquitoes are more numerous. Some insects have a damaging effect, like the locusts still found in Mediterranean areas, which can destroy large areas of vegetation. The pine processional moth destroys pine trees by burrowing into the trunk.

Birds like the golden oriole and black-headed bunting live on insects within the Mediterranean environment. Birds of prey, such as Eleanora's falcon, eagle, buzzard, vulture and kite, feed on insects and smaller birds. The pink flamingo, in the Camargue area of southern France, lives on water plants. There are gulls and shags, together with herons, egrets and shearwaters, throughout the Mediterranean. These are mainly fish-eating birds that find their food in coastal or freshwater sites.

Land animals such as the moufflon (horned sheep), wild boar, fox and bear live on plants or smaller animals. Hares, rabbits, shrews and mice are the most common small burrowing animals found in this region. The plants, insects, animals and birds are part of a delicate natural balance in the Mediterranean environment, forming individual ecosystems within the different types of habitat. If one part of the environment is changed, as with the destruction of trees, then the whole balance is changed.

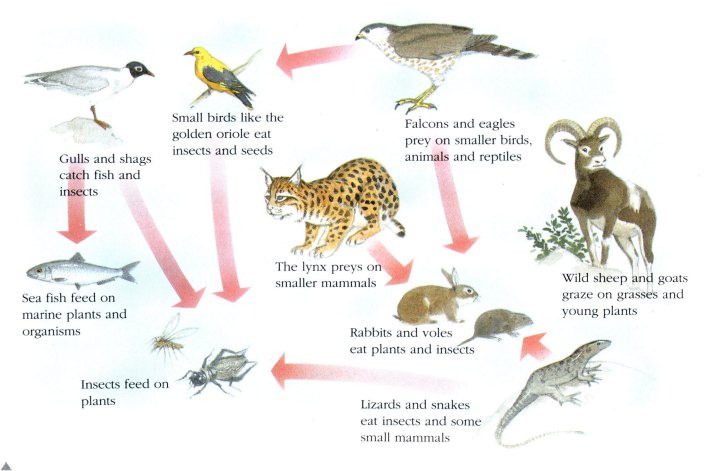

Gulls and shags catch fish and insects

Small birds like the golden oriole eat insects and seeds

Falcons and eagles prey on smaller birds, animals and reptiles

The lynx preys on smaller mammals

Wild sheep and goats graze on grasses and young plants

Sea fish feed on marine plants and organisms

Rabbits and voles eat plants and insects

Insects feed on plants

Lizards and snakes eat insects and some small mammals

▲
A food chain of some Mediterranean wildlife.

The red terra rossa soils colour the landscape of the Mediterranean. This picture shows olive trees planted in southern Spain. They are widely spaced to allow their roots to spread out and seek water.

MEDITERRANEAN SOILS

Soils in the Mediterranean areas are not very fertile. This is because the plants of the natural vegetation have few leaves, and those they have are leathery and decay very slowly. Plants supply little organic material to the soil, so in most areas farmers have to add fertilizers in order to grow crops.

The soils are red in colour and are called terra rossa soils (terra rossa is Italian for red earth). The colour comes from iron oxide left in the top layers of soil by rain-water as it

soaks down. The high iron content limits the variety of crops that can be grown.

Mediterranean soils have a lot of clay in them. In dry weather, the clay cracks and forms big, hard lumps. Farmers find these difficult to break up and in wet weather the clay becomes too heavy to work. Because they are so heavy, wet and rich in clay and iron, Mediterranean soils contain few soil creatures such as earthworms, which usually play an important role in helping to break up the soil and spread organic material.

· F A R M I N G ·

For centuries, farming in the lands around the Mediterranean was based on growing olives, vines and wheat. In addition, many farmers reared a few sheep and goats. In winter, the animals grazed on pastures near the coast. In summer they were herded up the hillsides to pastures where the cooler conditions gave better grass.

This agricultural system was well suited to the Mediterranean environment. The vine has long roots that can tap water from deep underground, the olive tree has narrow leaves and thick bark to reduce water loss. Both can survive in the shallow terra rossa soils. Olive trees were planted widely spaced to allow their roots to spread outwards. Wheat was sown in spring and harvested in June, before the start of the summer drought. Manure from the sheep and goats was used to fertilize the soils.

This pattern of farming did relatively little damage to the environment. The soil surface was protected from sun, wind and water by the trees, and the natural vegetation that grew alongside the farmland continued to support a wide variety of insects, reptiles, birds and animals. But the pattern began to change in the late nineteenth century, as the population living around the Mediterranean increased. This meant that trees had to be cut down to create new farmland, and timbers were taken for fuel and for construction. Farmers increased the numbers of animals they kept and the herds grazed on new growth in the grass and bushes, eating them away before full-sized plants could become established.

This destruction of trees, grass and bushes exposed the bare soil. The sun baked the soil, which became very dusty. The loose topsoil could be blown away by wind storms, which happened over large parts of Italy and Spain. In winter, heavy rainstorms washed away the fertile topsoil, especially in areas with steep slopes.

Once soil erosion began, the whole balance of the Mediterranean environment was upset. With the most fertile topsoil removed,

vines and olive trees died, and so did many of the natural-growing trees that had not been felled for timber. Other plants like the lupins, orchids, rosemary, myrtle and thyme struggled to survive, but many died.

As the range of plants decreased, so did the food supply and habitat for insects, birds and small animals. Fewer small creatures in turn meant less food for predatory birds and larger animals. Many species of plants and animals had declined drastically by the end of the nineteenth century, and the whole environment was slowly but surely being degraded.

◄

As the olive crop is harvested, plastic sheets are spread beneath the trees to catch falling olives.

▶

A market stall selling olives in Thessaloniki, Greece. Such stalls are a common sight throughout the Mediterranean, where olives and olive oil are important parts of the local diet.

Gypsies employed to bring in the grape harvest in southern Greece. Grapes are used for wine-making in many parts of Mediterranean Europe.

Motorized tools make the job much easier for farmers working difficult soils, like this stony ground near Mount Troodos, Cyprus.

MODERNIZED AGRICULTURE

Because the degradation of the Mediterranean environment occurred gradually over a long period, it was not until the 1950s that people began to realize their farming methods would have to change. Farms were often small and had little or no machinery. Harvesting was done by hand and wooden scratch ploughs were pulled by oxen or donkeys. Most farms produced just enough to feed the people who lived and worked on them – this is called subsistence farming. A few farms had surplus crops that they sold for cash.

Most people living around the Mediterranean were farmers or fishermen. There were few other jobs available for them, so farm wages were low. The people were poor and had large families, because the children were needed to work the land.

Since the 1950s, governments in Italy, France, Spain and Greece have encouraged changes in the patterns of farming in their countries. New crops such as oranges, lemons and grapefruits have been introduced, as well as vegetables like peppers, artichokes, asparagus, celery and avocados. These crops can be sold at higher prices than the olives, grapes and wheat of traditional farming, and are often grown for export. However, the fruit and vegetable crops need large amounts of water, so irrigation projects have been set up in southern France, Italy and Spain. Rivers have been dammed to create lakes, where water is stored for use throughout the hot, dry summer. Irrigation ditches carry the water to the fields where it is needed.

Small farms have been brought together into larger farms, making them more economical, and machinery has been introduced to do some of the hardest work. Farmers have been encouraged to join co-operatives, which allow them to buy their seed or fertilizer in bulk and at lower cost.

This new agriculture has brought prosperity to many farmers, but it has created different kinds of problems. The irrigated farms are on the best land, which is usually on the floor of river valleys. The steep hillsides have been neglected and soil erosion is still a problem there. The crops need lots of chemicals, especially fertilizers and pesticides. Not all of the chemicals are taken up by the plants, and natural water supplies have been contaminated. Chemical sprays have also killed many insects, reptiles and small animals. Hedges have been ripped out to create bigger fields, destroying habitat for all kinds of wildlife. Many species of plants, animals, reptiles and insects are under threat as the environmental system is changed.

FARMING IN THE WETLANDS

The modernization of Mediterranean farming has had important effects on the wetlands of the area. The term wetlands is applied to landscapes like swamps, marshes, estuaries or fens, in which water plays a key role. Wetlands cover six per cent of the earth's land area and form transitional zones between seas or lakes and dry land. They often seem to be inhospitable or unproductive, but the special environment is home to a unique collection of plants, insects and animals.

One such wetland area is the Camargue in southern France, the western part of the delta of the river Rhône. It has been built up over thousands of years from pebbles, sand and mud carried down the Rhône and deposited

◄

Traditional dishes in Mediterranean countries are based on ingredients available locally. This Spanish family is enjoying paella, which contains rice, chicken, seafood, vegetables and herbs, with side dishes of salad and olives.

▲
Combine harvesters collect the crop from a rice field in the Camargue.

where the river enters the Mediterranean Sea. The land is flat and covered by huge areas of reeds, one of the few types of plant that can survive in such wet, salty conditions. Elsewhere there are salt marshes and vast shallow lakes, called *étangs*. The largest lake in the Camargue covers 150 square kilometres but is only a metre deep.

The environment is home to millions of insects that breed in the warm, shallow water. In turn, hundreds of birds, including the golden oriole and the hoopoe, live there feeding on the insects. Heron fish in the shallow waters of the *étangs*, as do brightly coloured pink flamingos. Birds of prey like the osprey and even the rare lammergeier, a

A herd of the black bulls that roam freely in the wild landscape of the Camargue.

This aerial view of the Camargue shows the contrast of undrained wetland areas and orderly rice fields.

species of vulture, are found in the Camargue. Herds of wild white horses and black bulls move across parts of this unique wetland environment.

The ecosystem of the wetland has a precarious natural balance. If one element is changed, all of it may be damaged. For example, if the marshes are drained or the reed beds destroyed, then the birds, insects and animals die or are forced to move. Extensive alteration of the landscape most commonly means that wildlife populations die out; they can only move on and become re-established if there is a nearby site that remains unchanged.

Since the 1970s, the Camargue has been increasingly affected by agricultural changes. Southern France has a climate in which rice can grow well, and rice is a valuable crop that is in demand throughout Europe. But it has to be planted out in fields submerged to a depth of 10 cm of water for 72 days. Farmers have drained areas on the edges of the Camargue to create new farmland for

rice-growing. Once the land has been drained, the water supply to the fields can be carefully controlled.

France now grows all the rice it needs and has a surplus for export. However, the wetland ecosystem has been drastically altered by both the drainage and the chemicals used on the rice. The numbers of lizards, frogs, newts and birds living in the Camargue have all declined.

The Medjerda Delta

The delta is in northern Tunisia, at the mouth of the Medjerda river. It consists of gravel, sand and fertile silt carried down the river and deposited where the river meets the sea. Like the Camargue, it was once a wild, flat area covered by reeds, tall grasses, salt marshes and shallow lagoons.

In the 1960s, much of the delta was drained to create fertile farmland. The open landscape was transformed into an orderly system of fields and drainage ditches. The Medjerda is now a major producer of vegetables for the whole of Tunisia, but chemicals used to kill mosquitoes or fertilize the crops have polluted the ecosystem, destroying fish, birds and insect populations. The increase in agricultural production has been at the expense of the wetland environment.

▲
A generous selection of vegetables and fruits on display in a Tunisian market.

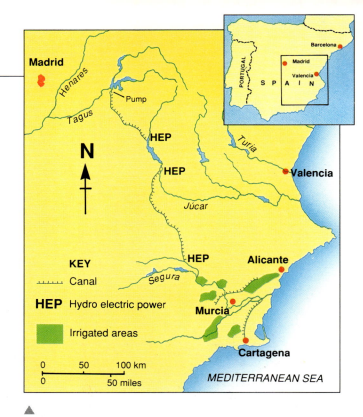

FARMING AND THE ENVIRONMENT

Irrigated farming is extremely important all around the Mediterranean, but especially in Valencia in south-eastern Spain. Here the summer drought is absolute and rivers like the Turia and Segura have little water. The area has over 3,000 hours of sunshine every year, which averages at eight hours per day. Plants can grow even in winter if water is available, so farming can be profitable.

To increase water supplies, water from the river Tagus, which flows westward, is diverted to the east. The water passes through tunnels and reservoirs until it reaches the river Segura, where it is used to irrigate the surrounding land. Cotton is grown on the irrigated farms, together with strawberries, peppers and calabrese. These food crops are transported by air to northern European cities such as London, Hamburg and Berlin.

Irrigation causes problems which affect both the natural environment and farming efficiency in the area. Water used to irrigate the land dissolves salts from the soil. Because conditions are so hot, some of the water evaporates and leaves behind the salts in the topsoil.

▲

This map shows the canals that divert water from the river Tagus to the Segura.

Over time, the soil becomes more and more salty – a process called salinization. Neither wildlife nor crops can survive and in some places farmland has had to be abandoned. The solution is to install field drains in the soil and to apply even more irrigation water. This flushes the salts out of the soil and the field drains carry them away into the rivers.

◄

Irrigation systems are often essential for efficient crop production on the dry Mediterranean soils, but they require vast amounts of water which may have to be channelled in from another region.

· T O U R I S M ·

▲
The seafront at the newly fashionable French resort of Nice, in November 1932.

The combination of sunshine, sea and ancient monuments draws about 100 million holiday-makers to the Mediterranean each year. Tourism has become a major source of income for most Mediterranean countries. In Spain, for example, the number of tourists grew from 1.3 million in 1952 to 48 million in 1992, earning the country some $14,000 million.

Tourism began in the nineteenth century, led by wealthy people from Britain, France and Germany. They began to spend their winters in southern Europe, seeing the milder climate of places like the south of France as beneficial to their health. Until the 1920s, seaside towns like Nice and Cannes were winter resorts. The visitors left before the heat of summer.

As sea bathing became more fashionable, from the 1930s visitors began to come in summer when the sea was warmer. However, the new visitors deserted the shingle beaches of Nice for the sandy shores of the Italian coastline, such as the beaches at Rimini on the Adriatic coast of Italy.

Until the 1950s and 1960s, most of the people from northern Europe who holidayed in the Mediterranean were relatively wealthy. Only they could afford the cost of travel and hotels, and also had enough free time to enjoy the Mediterranean. As a result, the tourists were few in number and their presence did not much affect the character of the environment in the places where they stayed.

However, at a few popular resorts such as Nice, Cannes, St Tropez and Venice, new hotels, restaurants and cafés were built. The sites with the best beaches and finest views along the coast were bought by developers, and sand dunes were levelled, marshes

The seafront at Alicante, Spain. Fast roads and high-rise buildings are now familiar features of the northern Mediterranean coastline.

drained, beaches created and new buildings constructed. In these sites, there was damage to the environment through loss of country-side and habitat, causing a drop in the local populations of insects, birds and animals. But tourism was not yet widespread along the Mediterranean coast and the damage was confined to a few, relatively small locations.

All this changed from the 1960s with the coming of package holidays and cheaper air travel. More people in northern European countries had jobs that provided an increased

Tourists waiting for a ferry in the Greek port of Naxos.

amount of paid leisure time, and preferred to spend their holidays in the sunny climate of the Mediterranean than in local resorts where the weather might be unreliable. New motor-ways made travel between northern and southern Europe much easier and quicker. Visitors from around the world came to enjoy the tourist centres and view the historic mon-uments. The result was a dramatic increase in the number of holidaymakers travelling to France, Italy, Spain and Greece.

TOURISM AND THE MEDITERRANEAN ENVIRONMENT

The growing popularity of the Mediterranean with increasing numbers of visitors has had a tremendous impact on the environment, mainly in coastal areas. More and more visi-tors need more and more hotels. Hotels are expected to offer instant access to beautiful sandy beaches and the warm Mediterranean sea, so there is great competition for sites along any stretch of coastline.

The pressures on the coastal environment have become immense. In Spain, for exam-ple, the coastal regions account for only 17 per cent of the country's land area, but 35 per cent of the Spanish people live there. Added to this, 82 per cent of visitors to Spain

stay on the coast. All the countries bordering the northern Mediterranean have seen massive coastal development – Spain, France, Italy, Greece and Turkey. Countries along the southern Mediterranean coast – Tunisia, Algeria, Morocco and Egypt – are all interested in developing their tourist industries. They have seen how Spain, Greece and Italy have become rich as a result of developing tourism, and it is a route that these poorer countries want to follow.

The competition for sites along the coast forces up the price of land. This means that local people can often no longer afford to buy farms or cottages in coastal areas. Only the commercial companies running the large hotels can afford the prices. Similarly, large-scale tourist development has swamped small traditional villages. In the 1950s, Benidorm was a small white-walled fishing village set on a headland separating two lovely beaches. Since then, the original village has been overshadowed by a mass of high-rise hotel blocks built directly behind the beach. Restaurants, cafés, self-catering apartments, bars, nightclubs and shopping malls have transformed Benidorm into a modern, popular resort. The local environment has suffered a massive loss of habitat, combined with pollution of sea and land by waste disposal.

Sponges on sale at the quayside on the Greek island of Rhodes. The souvenir trade is competitive; this trader has made a hand-written advertisement emphasizing the quality of his goods, and leaves his diving gear on display alongside the sponges.

CHANGES ON LAND AND SEA

Building hotels and whole new resorts along the coast completely changes the environment. Forests are cut down, sand dunes bull-dozed, marshes and wetland areas drained. New roads, water supplies and sewers are constructed. This massive loss of habitat has led to a decrease in the populations of plants, insects, reptiles and animals. In some places, even frogs and newts have become endangered species. In Spain, the lynx and brown bear, both once common around the Mediterranean, have been almost wiped out. New resorts are taking over beaches in Turkey, Greece, Tunisia and Morocco where the Mediterranean turtle used to breed, and this creature is now on the endangered species list.

Tourism affects the environment in other ways. Sponges and corals from the sea are being taken for sale as souvenirs for the tourists, so the marine environment is also being degraded. Even now, sewage generat-ed by coastal tourism is often pumped out to sea untreated. A World Bank survey has cal-culated that this applies to 70 per cent of the domestic waste from countries around the Mediterranean. Often the sewage is dis-charged through short outfall pipes close to the shore, and washes up on the beaches.

Raw sewage poisons the local sea bed, killing marine plants and fish, and even the birds that feed on them. For example, sewage has killed much of the Neptune grass, a type of seaweed, found along the coasts of Egypt and Libya. This neptune grass is an important habitat because it provides shelter and oxygen for young fish. As it dies, so do the fish.

Solid garbage is another problem. Plastic ropes, bottles, bags, sacks and packaging do not decay naturally. They are frequently dumped in the sea and are carried back to the coastline. For example, waste originating from Lebanon has been found on beaches in Cyprus and Turkey.

Hammamet-Nabeul

Hammamet–Nabeul is a new, purpose-built resort on the coast of Tunisia. It is built on a former wetland area, which consisted of sand dunes, marsh and shallow lakes. Now large modern hotels line the sandy beaches and behind them are shops, cafes and nightclubs.

Employment in Hammamet–Nabeul is growing as local people find work in hotels, shops and restaurants, but there is a price to pay for their increased prosperity. The fragile ecosystem of the sand-dune wetland area has been severely damaged by the new buildings and roads. The lakes and dunes were home to hundreds of species of birds, including egrets, herons, eagles and owls. These have died out through loss of habitat, or have moved away.

The area has been sprayed with chemicals to kill mosquitos, which annoy the tourists. However, the sprays also kill insects that are attractive and ecologically useful, such as moths.

▲ *Before development, the coastal landscape consists of sand dunes separating the beach from the low-lying wetlands.*

▲ *Drainage channels are dug across the wetlands, the dunes are flattened, and construction work begins.*

▲ *The site is transformed, with the wetlands converted to agricultural land and the former sand-dune area built over with hotels and leisure facilities.*

► *A typical example of the new hotel complexes at Hammamet-Nabeul.*

Greater numbers of people in the tourist resorts produce a vastly increased demand on water supply. In the hot, dry Mediterranean summers, water may be rationed, only available for a few hours in the morning and evening. Rivers have been drained of their water to supply the tourists, as have underground springs and wells. As a result, rivers and wetlands have dried out, plants have died and local wildlife has been forced to move away. In other places, reservoirs have been built to store the winter rains, so that water is available all year round, but even more wildlife habitat has been destroyed in the valleys that have been flooded to create the reservoirs.

PRESERVING A BALANCE

Lindos is a village on the Greek island of Rhodes in the eastern Mediterranean. In the 1960s it was a small village of whitewashed houses, with steep, narrow, cobbled streets. But it had a large, sheltered sandy bay attractive to holiday-makers in search of guaranteed sunshine and a clear, warm sea.

A holiday village has been built on the beach below old Lindos. The government has tried to make sure that tourism does not destroy the character of the old village, nor too much of the local environment. No hotels or other new buildings can be erected in the village, though some existing buildings have been modernized. The site of the beach development was restricted to ensure that local water and sewage facilities were not over-strained, and that destruction of sand-dune habitat was limited. As a result, local people make money from tourists by selling souvenirs from small shops in the streets, as

well as from bars and restaurants, but environmental damage has been minimized.

In Spain, near Valencia, new roads were built in the 1970s in advance of new hotels. Now the Spanish government has taken measures designed to prevent the coast becoming

Through careful planning and conservation, the village of Lindos with its ancient acropolis retains a picturesque quality, despite the fact that it is a busy tourist centre.

one long line of hotels from Barcelona to Benidorm. At Albufeira, a sand bar separates the sea from a freshwater lake, which is a stopping point for migrant birds like the marsh harrier. It is one of the most important ecological sites in southern Spain, protected by the government. The surrounding tarmac roads are being removed to return the area to its natural state. Through conservation projects such as this one, some of the eco-systems around the Mediterranean can be restored and protected.

The Library of Celsus is one of several interesting ruins at Ephesus, now in Turkey but formerly part of the ancient Greek empire. Sightseers studying the architectural detail need to move around freely and take a close view.

As the same site becomes crowded, people find it more difficult to view and walk around the ruins, and may get less enjoyment from their visit.

HONEYPOT SITES

Although the coast as a whole is the main attraction of the Mediterranean, there are particular places along the coast that are especially attractive. They may have beautiful beaches, or steep natural cliffs, or ancient monuments such as temples or Roman villas. These places are called honeypot sites, because tourists are attracted to them as bees are to honey.

Most visitors to honeypot sites expect to arrive there by coach or car, which can be parked nearby. Building new roads and creating ever larger car parks can quickly turn a beautiful, quiet area into a noisy eyesore. Many visitors to the Mediterranean come to see ancient sites such as Ephesus in Turkey, Knossos in Crete, or Delphi in Greece. Most of the tourists stay only an hour or two, but so that they can see the sites at their best, the old foundations have to be kept free of plants that would otherwise grow over them. The easiest way of keeping down plant growth is to use chemical weed-killers. These chemicals then spread across the whole site, killing plants, insects, birds and animals, and poisoning the soil.

Large numbers of people can destroy the very special features of such sites. Paths and buildings in places of historical interest are

being gradually worn away by the vast numbers of people scrambling over them. Too many feet trampling through sand dunes can destroy the marram grass and other plants that hold the dunes together – the sand can then be blown around by the wind and the whole site may change radically.

It is difficult to calculate the point at which honeypot sites become overcrowded. In theory, each place has a capacity, which refers to the number of people who can arrive without spoiling it. However, capacity can be judged in different ways. Physical capacity is simply the total number of people who can get on to a single site. Perceived capacity relates to people's ability to enjoy visiting a site that is becoming crowded, and this varies a lot from person to person. Ecological capacity means the number of people who can visit a site without causing environmental damage.

So far, no sites have been closed or formally restricted to prevent capacities from being exceeded, but it is something that may be taken into account in future tourist planning. It is being considered in Greece, for example, where some of the most popular tourist attractions are ancient monuments that have stood for hundreds of years, but are now in danger from erosion.

As with modern agriculture and tourism, industrial development has brought economic benefits to the people of the Mediterranean but at the cost of environmental damage to land and sea. Countries along the northern coast, like Spain, France and Italy, are relatively wealthy, and can now afford to consider how to deal with problems such as vanishing countryside and polluted sea. The countries along the southern edge of the Mediterranean, including Morocco, Algeria, Tunisia, Libya and Egypt, remain relatively poor but want to grow wealthier by developing their own industry and tourism. They feel that their first priority must be to raise living standards for their people, and concern for the environment can come later. They also point out that the richer countries continue to pollute the sea and destroy ecosystems, despite claims to the contrary. As a result, there is little agreement between the Mediterranean countries on how to conserve or improve their environment.

▲
An oil-spill barrier breaks up and is washed ashore coated with crude oil.

◄
In 1991, an oil spill from a tanker explosion threatens the Italian Riviera, the northwestern coastline of Italy. A floating barrier is laid out to stop the spread of oil.

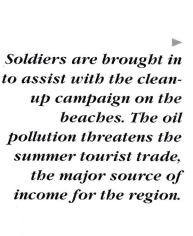

Soldiers are brought in to assist with the clean-up campaign on the beaches. The oil pollution threatens the summer tourist trade, the major source of income for the region.

THE OIL INDUSTRY

The Mediterranean forms part of one of the world's main oil routes, because it links the Middle East with the Atlantic Ocean via the Suez Canal, opened in 1869. As a result, large numbers of oil tankers pass through the Mediterranean Sea. The World Bank has calculated that at any one time there are about 2,000 merchant ships in the Mediterranean and of these 300 are oil tankers.

These ships carry water as ballast when they are empty, and the water becomes heavily polluted with oil. It is common practice for the ships regularly to discharge the ballast into the sea, and in this way about 600,000 tonnes of oil are pumped into the Mediterranean every year. As a result, beaches become polluted and fish and seabirds die, often covered in black, sticky oil.

Since the 1970s, there have been many conferences and statements (called protocols) aimed at stopping pollution of the Mediterranean, but some of the countries involved have disagreed with the protocols and refused to sign them. The Maritime Pollution Convention forbids the discharge of chemicals and hydrocarbons (oil is a hydrocarbon) from ships, but so far only nine countries have ratified (agreed to support) the convention. Half of the Mediterranean countries have not agreed to stop procedures such as dumping ballast, and many ships passing through the sea are registered outside the Mediterranean region and are not bound by the convention; so pollution continues.

Oil refining is another branch of the oil industry that causes environmental damage. Several large oil refineries have been built close to ports around the Mediterranean, especially in Spain, France and Italy. The industrial waste from these factories gets into rivers and into the Mediterranean itself. The Adriatic Sea between Italy and former Yugoslavia is particularly badly polluted, as is the sea near the industrial centres of Izmir and Istanbul in Turkey.

INDUSTRIAL POLLUTION

The growth of modern large-scale industries like mining, chemical production and electronic manufacturing is quite recent around the Mediterranean. However, factories have spread rapidly along the coast, expecially on the northern side. Some of these factories release waste containing poisonous metals such as mercury or cadmium, which build up in the sea and may be absorbed by fish. The fish may later be eaten by birds or humans, and in this way the poisons can enter the

food chain. Very high levels of poison have been found in the Adriatic Sea near Italy.

Pollution of the sea by oil, chemicals (from factories and farms) and sewage has led to a dramatic decline in the fish stocks of the Mediterranean. Fish such as swordfish, tuna and sardines have become scarce, especially in Corsica and Sardinia and off the coast of Turkey. The General Fisheries Council for the Mediterranean, part of the United Nations Food and Agricultural Organization (UNFAO), is trying to persuade all Mediterranean countries to reduce pollution and to cut down on fishing so that fish stocks can increase. Most countries have agreed to this, but action is very slow.

The European Community has started a series of fish farming projects around the Mediterranean, to increase the amount of protein available for people to eat. One such project is in the Gulf of Amvrakikos in Greece, an important wetland area consisting of lagoons, rivers and marshes. By 1993, six fish farms had been set up in the Gulf.

Although the fish farms provide work and food for local people, they also generate serious environmental problems. They take millions of gallons of water from the Gulf and pour back millions of gallons of waste, containing uneaten fish food and chemicals put in the water to prevent disease among the fish. This is upsetting the environmental balance of the water. The Gulf has traditionally been the habitat of over 300 species of birds, including a rare breeding colony of Dalmatian pelicans. The pelicans can only stay as long as the water remains unpolluted. As pollution levels rise, the birds die or fly away to find new nesting sites.

◄

Fishermen at Gabes in Tunisia prepare their boats against the background of a large industrial complex built along the shoreline.

Costs and Benefits of Industry

▲
Phosphate mines create a harsh landscape for the workers.

Algeria is making great efforts to develop its industries. Oil and natural gas from the south of the country are brought by pipeline to Algiers and Oran, from where some of it is exported and the rest goes to oil refineries and chemical factories. Phosphate mines south of Bougie are another basis of the chemical industry.

Most of Algeria's industry is located along the coast, especially near Algiers and Oran. The area has been transformed by the construction of new factories and blocks of flats for the workers, as well as schools and hospitals. But there has been serious pollution of the beaches and sea, especially from oil. Plants, birds, animals and insects have all died.

The Algerians are trying to cut down pollution from their factories, but they point out that much greater pollution still comes from the richer northern Mediterranean countries such as Italy, Greece, Turkey and Spain.

ENVIRONMENTAL CONTROL

The Environmental Programme for the Mediterranean (EPM) was started in 1988. It is funded by the World Bank and the European Investment Bank. Its three main aims are: to carry out studies to assess the problems of the area; to suggest programmes and policies to solve those problems; to provide funding and implement the projects that will improve the environment. By 1990, the studies aimed at defining the problems had been carried out, and work continues on developing and putting in place the projects designed to solve them.

In April 1990, after a big conference at Nicosia, in Cyprus, 16 countries signed the Nicosia Charter agreeing on the need to reduce pollution. The World Bank and European Commission agreed to invest US $1,500 million into projects for improving the marine environment.

All new EC initiatives to develop industry around the Mediterranean now have to include an Environmental Impact Assessment. This must calculate what types of pollution might result from the development and what needs to be done to ensure pollution is kept to a minimum.

Pollution of an Egyptian beach. In an effort to control such pollution, the Environmental Programme for Europe was started in 1988.

Timber Production in Portugal

In the Alentago area of central Portugal, where the typical Mediterranean climate extends inland, attempts are being made to establish new industry. This is an undeveloped, poverty-stricken region, with an acute need for money.

The European Community has funded a project to plant thousands of hectares of eucalyptus trees. The trees are fast-growing and are the basis for a new pulp and paper-making industry. The fibres from eucalyptus make good quality paper and local people can get better-paid jobs in the factories. In addition, Portugal will get a boost from the money it earns from paper exports. The trees have been called 'green oil' because they represent a new source of wealth for the area.

The benefits of the project are offset by a number of environmental problems. In order to plant the eucalyptus, traditional areas of holm oak and cork oak have been felled. This has destroyed habitat for rare animals like the lynx, and rarer birds such as

▲

A eucalyptus plantation near Redondo, Portugal.

Montague's harrier and the great bustard. The eucalyptus trees do not provide a habitat for these animals and birds and do not support the insects and small animals on which they feed.

Eucalyptus trees also dry out the soil, because they need so much water. The new plantations have

caused local springs and streams to dry up. The trees produce a thick blanket of waxy leaves, which cover the soil and are slow to decay. As a result, the soil gets fewer nutrients and soil animals die or move away. The whole ecosystem has been damaged by the eucalyptus.

·THE·GROWTH· ·OF·TOWNS·

*T*owns have existed around the Mediterranean for thousands of years. Most of the long-established towns are ports, like Brindisi in Italy and Piraeus in Greece, because the sea provided the easiest way to travel around. However, in the twentieth century many Mediterranean towns have expanded very quickly, partly because large numbers of people have moved to them from the

▲
Athens, the historic capital of Greece, has spread into a vast, crowded, modern city.

Pedestrian shopping precincts like this one in Seville are common to towns and cities throughout Europe. Depending on location, they may just supply the everyday needs of local people, or they may appeal directly to holiday visitors, for example by selling souvenirs and local crafts.

Flamingos in the Camargue seem undisturbed by the intrusion of a new housing development being built on the borders of this wetland area.

countryside. The number of farming jobs is falling: modern farming needs fewer workers. Jobs in cities are better paid; for example, a factory worker in Tunisia earns three times as much as a farm worker. In addition, housing conditions are better and there are services such as running water, sewage disposal, schools and hospitals.

In Greece and Turkey, many people have chosen to move to the biggest cities, Athens and Istanbul. Now a third of the total Greek population lives in Athens, more than 3 million people, and over 20 per cent of Turkey's population lives in Istanbul. The effects on the towns of this massive inflow of people have been dramatic. Huge new housing developments have had to be built quickly, simply to provide shelter for all the newcomers. Extra shops, schools, hospitals and clinics have also had to be constructed, together with lots of new roads.

All these developments have led to a loss of countryside. Large areas of land have been drained, reclaimed, bulldozed and built on. Habitats for birds, insects, animals and reptiles have been lost. Soils and farmland have been covered by tarmac, concrete and bricks. In the Camargue area of France, the town of Arles has become one of the fastest-growing points in the area. People and industries are moving into the town, attracted by its climate, fresh air and the availability of land for expansion. One result of this growth has been the drainage of large parts of the Camargue to provide space for houses, roads and new factories.

Tangier

The Moroccan town of Tangier is undergoing large-scale redevelopment geared to the tourist industry, similar to developments on the northern Mediterranean coast during the 1960s. On the seafront, adjoining sites have been bought up and the old buildings demolished, to be replaced by new hotels, apartments and offices. Public services and transport are modernized and efficient.

There are still extreme contrasts in the lifestyles of the people of Tangier. In a central district of the town, the houses of the very rich are lavish, enclosed by fences and private gardens.

On the outskirts, there are shanty towns with shacks built of waste materials. Poor people have come to the town from the countryside looking for work, but there is a high level of unemployment. Child labour is exploited and women and children are seen begging on the streets.

▲
This ornate house provides luxury and privacy for wealthy Tangier residents.

Elsewhere, signs of an improving, often Western-influenced standard of living are seen – imported consumer products, satellite dishes on the walls of private homes, and busy coffee houses where better-off local people spend their leisure time drinking and talking together.

▲
The poorest people of Tangier construct temporary homes from whatever waste materials can be found that will provide some shelter.

◄
New apartments like this are frequently too expensive for local people to buy or rent. They are likely to be let as accommodation for holiday visitors.

43

People living around the Mediterranean are becoming increasingly aware of the fragility of their environment. Regular reports detail the effects of pollution of the sea by oil tankers, industry or sewage. Similarly, the destruction of wetlands and farmlands is well documented, as towns, industry, agriculture and tourism all expand. The main problem is how to establish co-operation between all 18 Mediterranean countries in order to bring about environmental improvements.

It is difficult for the poorer countries of the south to concentrate their resources on the environment when their main priority is economic growth. They feel that such concerns will have to wait until both the country and its people are richer. Some countries lack the skilled workers and equipment required for cleaning up the environment.

Because of the flow of sea currents in the Mediterranean, a country may not see any benefit even if it does cut down on pollution. The benefits may be felt by a neighbouring country rather than by the people who have made the improvements. Some of the pollution in the sea comes from inland countries, carried along rivers like the Rhône, Po and Nile. Coastal countries argue that these inland nations should be made to pay something for the pollution they create.

There is currently a conflict in the Mediterranean area between economic growth and environmental protection. There are important measures being taken in three main

Basic development indicators - Mediterranean Countries.					
Country	Area (sq. km. x 1000)	Population (millions) mid-1990	Life expectancy at birth (years) 1990	GNP per capita (US$) 1990	Infant mortality (per 1000 live births) 1990
Spain	505	39.0	76	11,020	8
France	552	56.4	77	19,490	7
Italy	301	57.7	77	16,830	9
Yugoslavia	256	23.8	72	3,060	20
Albania	29	3.3	72	--- *	28
Greece	132	10.1	77	5,990	11
Turkey	779	56.1	67	1,630	60
Syria	185	12.4	66	1,000	43
Israel	21	4.7	76	10,920	10
Lebanon	10	2.7	65	--- *	--- *
Cyprus	9	0.7	77	8,020	--- *
Egypt	1,001	52.1	60	600	66
Libya	1,760	4.5	62	5,310	74
Tunisia	164	8.1	67	1,440	44
Algeria	2,382	25.1	65	2,060	67
Morocco	447	25.1	62	950	67

*statistics not available

Source of statistics: World Bank, *World Development Report*, 1992, Oxford University Press.

These figures show generally higher standards in the western European countries compared to those in North Africa and the Middle East, although Israel is an exception.

Unpolluted sea and clean, attractive beaches, like this one at Olu Deniz in Turkey, will ensure that tourists keep coming to the Mediterranean region and providing income for its people.

areas: protecting the land, by creating nature reserves and restricting development; saving wildlife, by maintaining habitat and setting up breeding programmes; and preserving the heritage of the region, with improvement and renovation of popular tourist sites like the city of Venice or the Acropolis in Athens, which have special environmental problems.

Undertaking conservation measures of this kind means giving up some of the economic benefits that tourism or industrial development might bring. But without care and protection of the environment, economic growth in coastal areas, particularly, is likely to decline. For example, in the early 1990s serious pollution in the Adriatic Sea led to a 30 per cent drop in tourism. People who live on the Mediterranean's shores – especially those of the poorer southern countries – have a right to a better standard of living. But if they try to achieve it without considering how new enterprises may add to existing environmental damage, it could be disastrous for both the region and its people.

GLOSSARY

Acropolis The stronghold of an ancient Greek city.

Ballast Any heavy material, such as sand, water or metal, carried on a ship to keep it stable.

Conflict of interests Disagreement over how important resources such as land, money or water should be used, caused by people having different concerns about their usage.

Conserve To protect something from harm and help to preserve it for the future.

Co-operative An organization of people with a common interest, such as farmers or factory workers, who can gain economic benefit from acting as one unit, for example, when buying food or supplies.

Culture A way of life which is passed on from one generation to the next. The culture of a region or nation can include many aspects of its people's social and religious customs, arts and pastimes.

Demand People wanting and being able to pay for something.

Drought A long period without rainfall, causing a water shortage.

Ecology The study of life forms in their natural environment and the relationships between them.

Ecosystem A community of plants, animals, insects and other organisms and the environment in which they live and react with each other.

Endangered species Particular types of plants, insects, animals, reptiles or birds which are in danger of becoming extinct (dying out completely).

Environment The surroundings of people, plants or animals.

Fertility (of a soil) The level of plant food in the soil, and its ability to sustain and nourish healthy growth in plants.

Hectare A unit of measurement equal to 10,000 square metres.

Habitat An environment which provides a home for plants, insects, animals and birds.

Honeypot site A very popular leisure spot which many people visit.

Income The money an individual or country receives in return for work.

Irrigation Watering the land to help crops grow.

Land use The purpose for which land is used; for example, for building on, or as farmland.

Nitrate A chemical fertilizer.

Package holiday A holiday where the flight and accommodation are arranged by a travel agent, usually for an all-in price.

Pesticide A chemical used to kill insect pests.

Pollution The presence in the environment of harmful substances such as chemical waste.

Pressures Forces to bring about change.

Recreation The things people do for relaxation and enjoyment when they are not working.

Soil erosion The natural removal of soil by wind or water.

Species A group of living things that are alike and which can reproduce.

Weather The day-to-day characteristics of the atmosphere, such as the presence of sun or rain.

Yield The amount produced by a crop, or the output produced per hectare of land.

BOOKS·TO·READ

Earthwatch by Penny Horton (BBC Books 1990)

Europe by David Waugh (Nelson 1985)

Europe Today by D. J. Davis and D. C. Flint (Bell 1987)

Europe and the Environment by David Flint (Wayland 1991)

Farming in Europe by David Flint (Wayland 1991)

Friends of the Earth Guide to Pollution by Brian Price (Maurice Temple Smith 1985)

Industry in Europe by Mark Smalley (Wayland 1991)

Living in Europe by David Flint (Wayland 1992)

Pollution and Conservation by Malcolm Penny (Wayland 1988)

Transport in Europe by Mark Smalley (Wayland 1991)

Tourism in Europe by David Flint (Wayland 1992)

Western Europe by Gordon Minshull (2nd edn. OUP 1989)

Western Europe by Neil Punnett (Basil Blackwell 1987)

·USEFUL·ADDRESSES·

C.E.A.T. (Co-ordination European des Amis de la Terre)
Rue Blanche 29
1050 Brussels
Belgium

Commission of the European Communities
20 Kensington Palace Gardens
London W8 4OL

Council for Europe
Boite Postale
431 R6
67006
Strasbourg
Cedet
France

Council for Environmental Education
School of Education
University of Reading
London Road
Reading RG1 5AQ

Earthscan
3 Endsleigh Street
London WC1H 0DD

Eurogeo
Henk Meyer
1Da Geographical Institute
State University
Heidelberglaan 2
P.O. Box 80115
3508TC
Utrecht
Netherland

Friends of the Earth (UK)
26–28 Underwood Street
London N1 7JQ

World Wide Fund for Nature
Panda House
Wayside Park
Godalming
Surrey GU7 1XR

INDEX